SECRETS OF
THE AMERICAN MUSEUM
OF NATURAL HISTORY

**WEIRD and
WONDERFUL FACTS
about
AMERICA'S NATURAL
HISTORY MUSEUM**

AILEEN WEINTRAUB

STERLING CHILDREN'S BOOKS
New York

STERLING CHILDREN'S BOOKS
New York

An Imprint of Sterling Publishing
1166 Avenue of the Americas
New York, NY 10036

ISBN 978-1-4549-2199-8

Distributed in Canada by Sterling Publishing Co., Inc.
c/o Canadian Manda Group, 664 Annette Street
Toronto, Ontario M6S 2C8, Canada
Distributed in the United Kingdom by GMC Distribution Services
Castle Place, 166 High Street, Lewes, East Sussex BN7 1XU, England
Distributed in Australia by NewSouth Books
University of New South Wales, Sydney, NSW 2052, Australia

For information about custom editions, special sales, and premium and corporate purchases, please contact Sterling Special Sales at 800-805-5489 or specialsales@sterlingpublishing.com.

Manufactured in China
Lot #:
2 4 6 8 10 9 7 5 3 1
12/18

sterlingpublishing.com

Photo credits on page 157

SECRETS OF
THE AMERICAN MUSEUM
OF NATURAL HISTORY

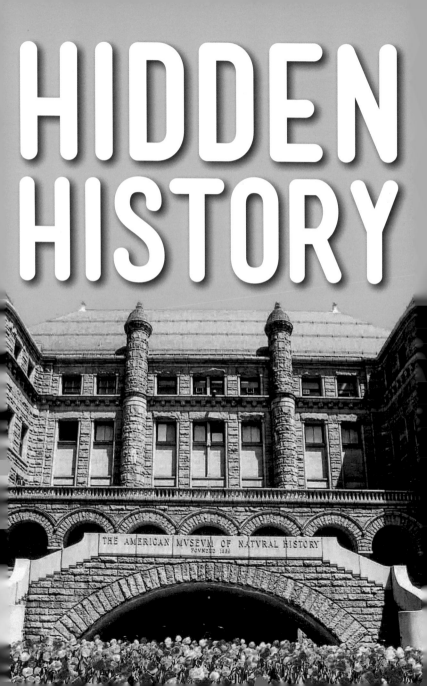

won't want to miss a trip through the Milky Way. These are just some of the things you can do at the American Museum of Natural History. There is something special here for everyone. People come from all over the world to visit this New York City landmark.

In *Secrets of the American Museum of Natural History,* we don't just tell you about the exhibits. We also take you behind the scenes to show you what really makes the Museum such an amazing place to visit. In this book, you'll find out about secret rooms, hidden treasures, and untold stories of the Museum's past. Then, when you visit the Museum, you'll have the inside scoop! Are *you* ready to find out the secrets of the Museum? Turn the page and let's get started!

INTRODUCTION

Imagine coming face-to-face with one of the biggest dinosaurs to ever walk the earth. Or maybe you'd rather relax under the belly of a blue whale. If space travel is more your speed, you

CONTENTS

The Central Park Arsenal was the very first home of the American Museum of Natural History in 1869. The top two stories of this building were used to show exhibits until a permanent space could be found. The building is very cool; it looks just like a **medieval castle**. Even though the Museum no longer uses it, the building is something you can still check out; it's now the headquarters for the Central Park Zoo.

The Museum opened on **77th Street** near **Central Park West** in **1877**. At that time, it was only one building. The collections grew and grew and are still growing today! Now, the Museum is actually **25** interconnected buildings! The **Northwest Coast Hall** is in the oldest building of the Museum.

Today, the Museum is so big, it takes up **four** city blocks! Each interconnected building was built at a different time. Can you tell when you move from one building to another?

There was no **electricity** when the Museum first opened, which made it hard to see the exhibits. So the Museum came up with a bright idea. They built **T-shaped display cases**. They put the cases against long, narrow windows. The light from the windows lit up the cases. You can see **12** of these windows in the **Hall of Vertebrate Origins**. Light only shines through **six** of them, though. The other **six** were blocked off when a new building was added.

There was a plan to make a **half-mile walkway** called the **Inter-Museum Promenade** that would connect the Museum with the Metropolitan Museum of Art. It never happened though. This plan would have meant big changes to Central Park. Instead, visitors walking between the two museums still use the beautiful winding paths that run through the park. Try it one day. There's so much nature, you'll forget you're in a big city!

Approximately **5 million** people visit the Museum each year. How does the Museum keep everything up and running?

The Museum has:

Over **1,000** employees

★ ★ ★

More than **200** scientists

★ ★ ★

500,000 square feet of public space

★ ★ ★

45 exhibit halls

★ ★ ★

27 elevators

The artists who put together the **dioramas** like to have **fun** with their art. Sometimes they add something that's not supposed to be there. In the gorilla diorama, the artist painted **gorillas** behind the grass and trees. No one will ever see these gorillas, but the artist put them there anyway.

Two species of *Phyllobates*, the poison-dart frogs used by Chocó Indians of the Emberá and No-anama groups. Toxic skin secretions make these frogs distasteful to most predators. See Case 7 for structures of the toxins.

Poisonous darts

collected from South America are hidden deep inside the Museum. They are locked in a **very special safe to keep them secure**. These darts were used to hunt animals with poison collected from poison dart frogs (see page 57). Now, they are part of the Museum Library's Memorabilia Collection.

How tall is the biggest tree you have ever seen? In the **Hall of North American Forests**, you can see a slice of a tree that was over *300 feet* tall. In 1891, lumberjacks cut down this **giant sequoia** tree that was over 1,400 years old. It is now illegal to cut down giant sequoias. These incredible trees have **fire-resistant bark** and don't often get diseases. Giant sequoias can live for more than 3,500 years!

On the fourth floor is a staircase right next to the Titanosaur's head. Look up and you'll see two large metal rings at the top of the stairwell. Before the Museum had elevators, workers attached thick ropes and chains to the rings. That's how they pulled exhibits from one floor to the next!

Not afraid of **bugs,** are you? Walk through the **Hall of North American Forests** to see a mosquito **75** times the size of a real one! The glass, wax, and wire model was made in **1917** to teach visitors about the dangerous diseases some mosquitoes carry.

Have you ever been given a nickname for being really good at something? Well, Alice Gray had. She was known as the **"Bug Lady."** Why? She was really good at drawing bugs! Alice began working at the Museum in 1937. It didn't take long until she was in charge of all the **live insects** and **spiders**. Gray loved taking bugs to New York City schools to teach kids about them. She even took bugs on *The Tonight Show.*

A mineral called the **Subway Garnet** sounds like it was found in the New York City subway. But this **9-pound** mineral was actually found in a sewer in **1885** when New York City didn't even have subways! It was dug up from deep beneath Manhattan, so it became known as the Subway Garnet. This **430-million-year-old** mineral is almost the size of a **bowling ball**. The Subway Garnet is currently kept out of sight in the Museum's collection.

Did you know that one of the world's finest mineral specimens was once used to pay a bill? It's true! In 1952, a miner used the **Newmont Azurite** to pay an outstanding tab. This mineral is famous for its large size and dark, **nearly perfect crystals**.

The Newmont Azurite was formed when rainwater and groundwater joined together on copper ores. To the naked eye this mineral looks black, but it's really a dark blue. If you're wondering why this beautiful azurite wasn't made into jewelry, it's because the mineral is too soft and fragile.

Want to explore natural history? There are four different ways to see objects from the Museum!

1. Visit the permanent exhibits at the Museum

2. See temporary exhibits that travel to museums all over the world

3. Check them out online at amnh.org

4. Read books and magazines at your library that show the Museum's amazing collection

QUICK
QUIZ

World-famous magician Harry Houdini made a donation to the Museum. Can you guess what it was?

a) his stuffed gray parrot

b) a book of magic tricks

c) a tiger

d) his handcuffs

Answer: a

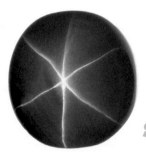

The Museum is also home to the **Star of India.** It's the world's largest gem-quality blue star sapphire. Not only is it beautiful, but it's also one of a kind. A special mineral gives it a milky quality and starlike effect. And if you think that's not cool enough, this amazing sapphire is about **2 billion years old!**

The water-clear quartz known as rock crystal was believed to be what?

a) glass eyes

b) permanently frozen water

c) crystal balls

d) both b and c

Answer: d

Can you guess what the Museum's biggest collection is? If you guessed **insects**, you'd be right! The Museum is home to **17 million** insect specimens. Because there are so many, they aren't all on view. The Museum also has the largest and most diverse collection of **spiders** in the world.

Mud and Hermes are African spur-thighed tortoises that were kept as pets in the Herpetology Department. That's where scientists study amphibians and reptiles. This type of tortoise is the third largest in the world. Each one can weigh up to 250 pounds! Mud and Hermes retired with their keeper, but you used to be able to spot them taking a stroll on the Museum's lawn!

A famous movie called ***Night at the Museum*** takes place in the American Museum of Natural History. Some scenes were **filmed right outside the Museum**, but most of the movie was filmed in Canada! Directors took lots of notes and photographs so they could build a set that looked just like the Museum. The movie may not have been filmed at the Museum, but it certainly got more people to visit!

Each year, the Museum creates several special exhibitions that premiere at the Museum, after which many of these exhibitions then travel to other museums around the world. These exhibitions have been seen on six continents—all but Antarctica.

QUICK
QUIZ

What is the word scientists use to describe samples they collect?

a) specimens

b) models

c) cases

d) expeditions

Answer: a

The "Golden Corridor" is a **hallway** on the **fifth** floor of the Museum. It's nearly the length of **1 city block**. This hallway is filled with large cabinets. These cabinets hold some of the Museum's **34 million** specimens and artifacts. These include **invertebrates** and **small mammals**. The Golden Corridor is the longest hallway in the whole city.

Most people don't know that the Museum has its very own **library**. With **16 miles of books and collections**, it's pretty big. That's longer than the entire island of Manhattan from end to end! The books are packed into a **7-story** building along with photographs, films, videos, archives, art, and Museum memorabilia.

Hidden in the main library is a **Rare Book Collection**. The **14,000** priceless books in this special library are so important, they are kept under lock and key. Some are over **500 years old**!

The **largest book** in the Rare Book Collection is called *The Birds of America.* It's over **3 feet** tall and takes **two** people to safely pull it off its shelf. This nearly **200-year-old** book is worth **millions** of dollars! That makes it the **most valuable** book in the whole Museum.

Many of the books in the
Rare Book Collection were bound
by hand. A few books in the collection
were even handwritten, and some of the
illustrations were colored by hand.

What's the oldest book
in the library called?
De Animalibus

★ ★ ★

What is it about?
Animals

★ ★ ★

When was it published?
**The year 1495 (over 500
years ago!)**

★ ★ ★

Who wrote it?
A man named
Albertus Magnus

Many specimens and artifacts used to be **hidden away** in the attic of the Museum. Lots of them were tucked away in corners, and it could take **hours** to find a single object! In the **1980s**, Museum experts began sorting. Then, they started to put the information into a computer database. Many objects can now be found in **seconds**!

There are **hidden elevators** throughout the Museum. The largest one is **17 feet** long and **11 feet** wide. It's mostly used to move big exhibits. If you visit on a crowded day, you might be lucky enough to **catch a ride**, because sometimes it's used as an express elevator for visitors from the first floor to the fourth floor. Look for it on the first floor between the **Hall of North American Forests** and the **Hall of Biodiversity**.

The **Frick** building has the most floors in the Museum. It is **11** stories high. The building is **invisible** from outside of the Museum grounds. That's because it's hidden in one of the courtyards. The Frick is where they store **fossilized bones**. The bones are so **heavy** that the Museum had to add special steel beams to hold up the building!

There are **mysterious doors** on many of the stairway landings in the Museum. These doors are always **locked**. Wondering what's on the other side? Other buildings! The Museum was **built in stages**. That means the staircase landings of some buildings connect to a whole other floor in the next building. This is why the ceilings throughout the Museum are different heights.

Many **secret rooms** in the Museum are hidden in plain sight. Some are storage rooms filled with ancient artifacts, and some are offices. Next time you are in the **Ross Hall of Meteorites**, if you come in through the **Spitzer Hall of Human Origins**, look for a secret room on the left. Can you find it?

People from all over the world visit the Museum. So, at the end of the day, the Museum's message, "The Museum is now closing," is announced in 10 languages: English, French, German, Russian, Chinese, Spanish, Portuguese, Italian, Japanese, and Korean.

What's the longest amount of time you've ever spent vacuuming? Ten minutes? Twenty? It takes two whole days to clean the dust from the **blue whale** on display in the **Milstein Hall of Ocean Life**. The massive whale, made of fiberglass and polyurethane, is cleaned once a year. One person rides in a cherry picker to do the job, and it's often streamed live on the Internet! This whale has been on display since 1969.

On the Wednesday before Thanksgiving, the blocks around the Museum close down. **Gigantic balloons** take over the streets for the **Thanksgiving Day Parade**! More than **one million** people come to see the action. All of the balloons spend the night outside the Museum, where they're placed under **huge nets** to keep them from floating away. The next day, the parade starts at the Museum.

Do you dream of going on expeditions and decoding the secrets of the universe? About 200 scientists work at the Museum doing just that. These men and women work hard to bring new discoveries to the Museum and its many visitors.

WILD
ADVENTURES

In the early **1900s**, the Museum hired **explorers** to travel to South America, Asia, and Africa. These brave men and women headed to **hidden forests**, unmapped deserts, **unknown swamps**, and **unexplored coasts**. When they found something interesting, they took it back to the Museum for further study.

Collecting the bones of large mammals was stinky work! In **1907**, Roy Chapman Andrews and his team collected a **whale skeleton** in New York Harbor. It was **freezing cold** outside. Strong winds buried half the skeleton in the sand. The team had to clean the bones before they could bring them back to the Museum. By the time they were finished, they smelled awful!

Museum scientist Frank Lutz and his team found some interesting ways to collect insects in Cuba. They would **tap trees with umbrellas** to get the insects to jump off! Sometimes, they used **giant nets** to sweep the grass. Other times, they broke up rotting logs or dug in the sand. They even chased after the flying insects!

Did you know there are more living **insects** than any other group of organisms? In fact, scientists estimate that there are around **10 quintillion insects** living on Earth. That's 10,000,000,000,000,000,000 insects! The Museum collections include **7.5 million gall wasps** and **3.5 million moths and butterflies**, among many other types of insects.

During the early years, the Museum wanted to increase its insect collection. A man named William Beutenmüller made **six trips** to the Black Mountains of North Carolina. He collected insects there that had never been found anywhere else in the entire world. By the time he was done with his expeditions, he had collected more than **30,000 specimens!**

Roy Waldo Miner and his team came up with a way to study coral reefs under water. First, they had to dive down 30 feet to a special viewing chamber. The chamber was attached to a tube that pumped in fresh air from the surface. Here, they could sit and draw pictures of the reef. Eventually, they re-created the reef in the Museum. It took 12 years to get it just right! You can see the coral reef in the Milstein Hall of Ocean Life.

QUICK QUIZ

Someone who decides what is put on display in a museum is called a:

a) preparator

b) maintainer

c) curator

d) helper

Imagine studying **the world's most powerful lizard** up close. That's exactly what scientists did in **1926**. They traveled to Indonesia, in southeast Asia, to visit the faraway island of Komodo. Here, they came face-to-face with the **10-foot-long, 200-pound** lizard. *Yikes!* Check out the **Hall of Reptiles and Amphibians**. There, you'll see a **Komodo dragon** about to eat a wild boar for lunch!

Searching for **poisonous frogs** sounds dangerous. Most people wouldn't want to spend time with poisonous creatures, but Charles Myers did just that and made an **amazing discovery**: babies born to poisonous frogs at the Museum weren't poisonous! The frogs were only poisonous in the wild. So, where did the poison come from? The poison came from the insects the frogs were eating in the forest.

In **1926**, a group of paleontologists on an expedition in New Mexico found the **tip of a spear** in the ribcage of a **10,000-year-old** bison. This important discovery proved that humans had lived in North America at least **10,000** years ago! You can see the spear tip on the third floor of the Museum. It's near the **Hall of Plains Indians**.

When scientists go on an expedition, they don't just start digging in the dirt. Instead, they make sure they have a plan before they get to work. Here's what they do in the field.

Ask a lot of questions

Study maps

Look for clues

Make observations

Record what they see

Talk to other scientists about what they find

Share their discoveries and publish them

Looking for fossils is like putting together pieces of a puzzle. On one expedition, scientists found the **teeth of ancient apes** in a cave in Vietnam. Porcupines had chewed most of the teeth. Luckily, the scientists still found **important clues**. They could tell that some of the teeth belonged to an ape as big as **10 feet tall**. These teeth were twice as wide as a modern-day gorilla's teeth!

Word Search

Let's dig up some fossils. Find the tools you'll need. (Words can be found vertically, horizontally, and diagonally.)

```
R C B A C W C V J F H H A I Z
P I Z J N V A G U Y M O A V J
O S F E L J M N M C Q D X K J
S H O V E L E G O B K P B Z G
G V X B H R R U O D C K F J H
M O B I I E A O S X D P L E J
Z E L D Z P T E H S X X F Z F
U L X U F S H X O E Y R D C U
G L O V E S A Y Z F X S L G G
N U M Z X K M X P I H S Y R P
H A V P N U M T S N U N F P I
P H C A V L E Q Z K O L J X T
H C F K X Q R J T G E S B N B
F G Z S E E T K W B P X S Q O
Z P P B P R C X I W T N B G X
```

axe   knife   gloves

hammer   camera   shovel

boots   map

Most of the time when a scientist finds a shark fossil, it's been crushed to pieces. That's because shark cartilage is made up of millions of tiny "tiles" called tesserae that usually fall apart unless the skeleton gets buried very quickly without being disturbed. When scientists found a perfect 325-million-year-old shark fossil, they knew they had something special. So how did they look inside the fossil? They took X-rays!

In the **mid-1900s**, Robert Kane was exploring in Africa. It was his job to make a mold of a **dangerous viper snake**. The viper was already dead, but Kane was still nervous. When the mold hardened, Kane removed the snake. That's when the snake grabbed for Kane's hand. The viper wasn't dead after all! Luckily, it just missed biting him. You can see this very same viper in the mandrill diorama in the **Akeley Hall of African Mammals**.

On Rapa Nui, also known as Easter Island, 887 moai exist. Moai are huge volcanic rock carvings of ancient people. Many islanders consider the carvings to be sacred. So it is really special that the American Museum of Natural History has a moai cast. During an expedition to Rapa Nui, in 1934–1935, a mold was made of a moai. Then, a plaster cast was created so this incredible object could be put on display in the **Hall of Pacific Peoples**.

QUICK
QUIZ

On average, how many scientific expeditions does the Museum have each year?

a) ten

b) fifty

c) one hundred

d) thousands

Answer: c

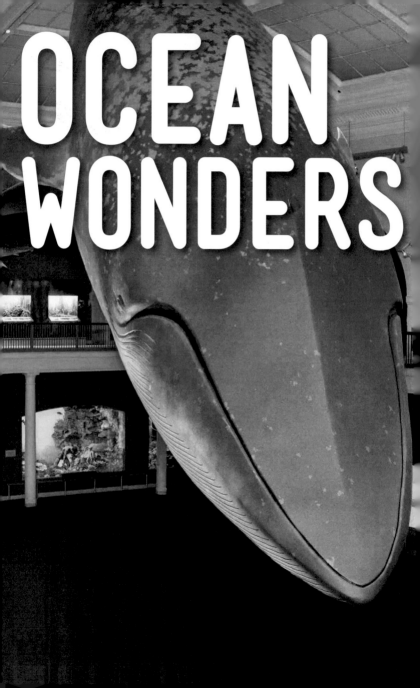

OCEAN
WONDERS

The **whale** you see today in the **Milstein Hall of Ocean Life** wasn't the first life-size whale in the Museum. Roy Chapman Andrews built the first whale in **1907**. He used an iron frame, netting, wood, plaster, and papier-mâché. It hung in the **Hall of Mammals**, since whales are mammals, for more than **50** years.

Wondering how the **blue whale** stays suspended in midair in the **Milstein Hall of Ocean Life?** Many people think that it hangs by strings, but that's not true. The whale has an interior frame made from iron, which connects to a large steel pipe that extends up into the roof.

It's not easy to hang a giant whale from the ceiling. It took **a long time** to figure out the best way to do it. Here are some of the ideas the Museum came up with that didn't work.

1. Hang the whale by wires.

2. Make the whale out of rubber and fill it with helium, like a giant balloon.

3. Put the whale on a platform and hold it up with a giant rod.

The person in charge of making the whale was worried that it would fall. Every day during construction, he used a **rod** to measure from the floor to the whale's **chin** to make sure the whale didn't droop. On opening day, the rod was once again placed between the whale's chin and the floor. Guests thought it had to stay there to hold up the whale. Suddenly, the Museum president pulled the rod away. **People were amazed** when the whale stayed up on its own.

Today, the blue whale is one of the most popular museum exhibits in the entire world.

Length:
94 feet

★ ★ ★

Weight:
21,000 pounds

★ ★ ★

Made of:
Fiberglass
(a lightweight material made from thin threads of glass)

★ ★ ★

Amount of paint used:
25 gallons

When you visit the whale, lie down underneath it and look up. Try to spot the whale's **belly button**. It wasn't always there! In **2003**, the Museum made changes to the whale. The belly button was added to make the whale more true to life. The Museum also painted the whale and reshaped the tail and blowholes.

This **29,000-square-foot** hall is a must-see stop on your visit. Find it on the first floor!

It has:

Computerized, glowing **jellyfish**

★ ★ ★

750 models of sea creatures

★ ★ ★

14 dioramas

★ ★ ★

Touchable models of a **pearl oyster** and a walrus tusk

Sometimes fish specimens are kept in jars. These glass containers are filled with a liquid that keeps the fish from rotting. But visitors complained that they couldn't really see how beautiful the fish were. The Museum came up with a solution. They began making **plaster molds of fish**. Then, they covered the molds with **fish scales** and painted them. This made the fish look more lifelike and colorful.

Today, you can see many beautiful models of fish in the **Milstein Hall of Ocean Life**. The newer models are made from plastic and a lightweight material called fiberglass. That was not always so. Originally, models were made from painted **beeswax**!

Secret storage rooms filled with animal specimens, such as fish, are hidden out of sight. The Museum has over **2 million** fish specimens! Many fish specimens are are grouped by gender, birthplace, and age. They are also grouped by type. This allows scientists to study and compare all the different fish.

QUICK **QUIZ**

What do you call the study of fish?

a) paleontology

b) ichthyology

c) lepidopterology

d) geology

Answer: b

Most people think the whale shark is a whale. It's really just a big fish—a whale-size fish! It happens to be the biggest fish in the world! The largest whale sharks can grow up to 40 feet in length and weigh up to 20 tons! Divers like to swim with these gentle giants in the ocean. They feed on plankton and aren't a threat to humans. You can see one on the upper floor of the Milstein Hall of Ocean Life.

The model of the **Andros Barrier Reef** in the **Milstein Hall of Ocean Life** is **2 stories** high and weighs **80,000** pounds. To build it, workers first coated the reef with beeswax to make it look alive. Then they built a **hidden frame** to support it. They used **ropes**, **pulleys**, **drills**, and **paint** during the diorama's construction. Next, they made model fish and polyps, which are small sea animals, from blown glass and wax.

It wasn't easy putting together the **Sperm Whale** and **Giant Squid** exhibit. A full-size model of the whale was too big to fit in the diorama! The Museum solved the problem by putting only part of the whale in the scene. We see the head of the whale and part of the squid in the dark, deep sea.

Giant squids can grow up to **60 feet** long! They live **deep in the ocean** where there is no light. That's why the squid and the whale diorama is so dark. At first, both animals were painted black, but visitors couldn't see them! They thought the diorama was empty. The Museum added a **blue light** and repainted the animals.

The giant squid you see on display is a model. But the Museum has a **real giant squid**, too. It's **25 feet** long! It was caught in a fisherman's net by accident off the coast of New Zealand. The squid was sent to the Museum **frozen**. It is now stored in a **large steel tank** in a special storage room.

Dinosaurs aren't the only animals that became extinct **65 million** years ago. **Ammonites** are ancient creatures that lived beneath the sea in coil-shaped shells. They lived in the ocean for **300 million** years before being wiped out. That means they were around way longer than dinosaurs! Luckily, we can still study their fossils. Check out the **75-million-year-old** ammonite in the **Grand Gallery**. Its shell looks like a rainbow.

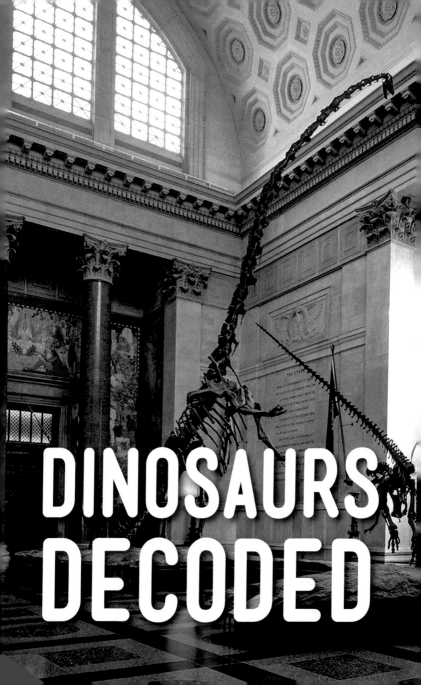

DINOSAURS
DECODED

Walk in the **main entrance** of the Museum and step back in time. **Prehistoric time,** that is. A giant **Barosaurus** is protecting its young from an attacking **Allosaurus**. The *Barosaurus* stands straight up, ready to fight. The Museum is one of the few places that have a *Barosaurus* cast on view. This *Barosaurus* is the **tallest dinosaur** mount in the whole world.

The **first dinosaur fossil** ever collected for the Museum was the pelvis of a *Diplodocus*. In 1897, fossil hunters Henry Fairfield Osborn and Barnum Brown discovered this **long-necked** dinosaur in Wyoming. The *Diplodocus* is **80 feet long!** That's the length of **five** cars lined up end to end! Find this fossil on display in the Hall of Saurischian Dinosaurs.

Most people think that scientists **dig out dinosaur bones** in the field. That's not true! When a dinosaur fossil is discovered, scientists **dig a trench** around it. Then they wrap the whole thing, soil and all, in bandages soaked in plaster. They take the whole piece back to the Museum, where they dig out the bones. Scientists use a variety of tools, like **chisels**, **hammers**, **needles**, and **brushes**. It can take **over a year** to dig out **one** single fossil!

The **Titanosaur** on the fourth floor is a cast of one of the **largest dinosaurs ever discovered**. It's **122 feet** long! The Museum could not find a space big enough for it, so they decided that its head would stick out into the hallway!

When you get off the 77th Street elevators on the fourth floor of the Museum, watch out! The giant

TITANOSAUR

will be waiting to greet you.

It lived about **100 million** years ago.

★ ★ ★

It has a thigh bone that is **8 feet** long, which is taller than the tallest basketball player.

★ ★ ★

Its neck is **39 feet** long. That means it could easily have peeked into the window of a 5-story building.

★ ★ ★

Its remains were discovered in Argentina in **2014**.

When living, this
Titanosaur would have
weighed 70 tons, or
as much as five school
buses. This big dino was an
herbivore. That means
it only ate plants . . . *lots* of
plants!

QUICK QUIZ

Scientists now know that some dinosaurs had feathers. Which animal are dinosaurs related to?

a) skunks

b) cats

c) birds

d) elephants

Answer: c

Have you ever wondered where dinosaurs get their names? Scientists have used Greek and Latin to name all their scientific discoveries for hundreds of years. Dinosaur names often describe special features of fossils, like "spiked" or "lumpy." Sometimes dinosaurs are named after the paleontologists who discover them, or the places where they were found. If you discovered a new dinosaur, what would you call it?

There's an interesting **weather vane** on the top of the Museum's **Power House Building**. What makes this weather vane so special? It's shaped like a **Stegosaurus**!
To find it, head over to the **Arthur Ross Terrace**. Turn your back to the planetarium and scan the rooftop of the building in front of you.

In the early 1900s, scientists didn't think dinosaurs were very fast. Then they discovered the Ornitholestes dinosaur. This was a fast-moving predator. Learning about this dinosaur changed everything! It challenged a lot of what scientists thought they knew about these prehistoric creatures.

QUICK **QUIZ**

The plant-eating *Camarasaurus* has a thigh bone that weighs *650 pounds*. What's the name of the room where they store this and other giant bones?

a) The Little Fossil Room

b) The Big Bone Room

c) The Dinosaur Room

d) The Giant Fossil Room

Answer: b. The Big Bone Room has the biggest bones in the Museum. This secret room is not open to visitors.

Be sure to visit the **T. rex** in the **Hall of Saurischian Dinosaurs** on the fourth floor. When scientists tried to put it together, its head was too fragile and **too heavy** to be on display! Instead, they made a replica. The real skull is in a case below.

Ever hear of a **dinosaur mummy**? It sounds like science fiction, but you'll find one in the **Hall of Ornithischian Dinosaurs** on the fourth floor. The mummy is a fossilized **duck-billed dinosaur's skin**. When it was unearthed, scientists could see a clear impression of its skin in the stone. This mummy is one of the **most complete nonskeletal dinosaur remains ever found**.

Check out **real-life dinosaur footprints** on the fourth floor. In **1938**, **107-million-year-old** fossil footprints were dug up from a river in Texas. They are known as the **Glen Rose Trackway**. There are **two** sets of prints. The small ones are from a dinosaur that walked on its two back legs. The other set is from a larger, plant-eating dinosaur that walked on all four legs.

The **Stegosaurus** is a great **big** dinosaur with a **tiny** head. Some scientists used to think it had a **second brain** because the one in its head was so small. We now know that the *Stegosaurus* had only one brain after all. Look for the *Stegosaurus* on the fourth floor.

Word Search

Search for the dinosaur words below.
(Words can be found vertically,
horizontally, and diagonally.)

```
T V J G Y Z Z D F Q W M K K S
R Y D W I X L J V R L P P I Z
A U Q S N E I D Z Z B S N G E
X J J A E R M K X U E I L F S
V Q Y Y G W O B F S X W F R T
S C Y L U S X H L M V B E B C
P R E H I S T O R I C H O O N
U J K C S E T X Y T T I S N I
G V G H M P U K I A U G H E T
D G Q H X R G U E S U S F E X
K D O E C L Y F P Z C L U Q E
R H G L A Z Z Q J Y E X G T T
L G A Y S V T U J T N B Y Y I
S W T Y A G B L I S S O F B Y
S A K G U E R G P S Q C F Q P
```

bone   fossil   extinct

feathers   eggs   prehistoric

claws   horn

The Museum's **Triceratops** is **65 million** years old. It has **three** horns. Two are over its eyes and one is over its nose. These dinosaurs are more like us than you might think: the *Triceratops* horns are made of a material very much like our **fingernails**!

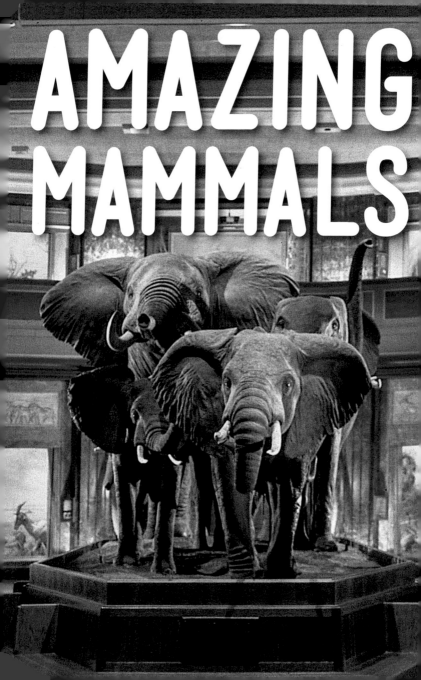

AMAZING
MAMMALS

The **Akeley Hall of African Mammals** was named after explorer, artist, inventor, and conservationist Carl Akeley. In **1909**, he decided to create scenes in the Museum that looked like natural habitats. These scenes are called **dioramas**. Akeley spent more than **17 years** designing, planning, and raising money for the African mammal dioramas.

Head to the **Akeley Hall of African Mammals** to look for elephants, lions, gorillas, and rhinos! Each scene in this hall comes from nature.

Scientists spent a lot of time in the field studying the animals and environments. They used their observations to create similar scenes at the Museum. Look closely. The lighting in each diorama shows a specific time of day!

The Museum dioramas have remained largely unchanged for years. That's because they are sealed behind glass. The only way to get in is by removing the large viewing glass. This glass is very heavy and hard to move. That means that someone only goes into a diorama if repairs are needed. Once a diorama is closed, no one goes in again for a very long time!

Word Search

Search for the animals in the
Akeley Hall of African Mammals.
(Words can be found vertically,
horizontally, and diagonally.)

```
T O S T R I C H A F O O D S Y
J N Z K J U A N H K K G R W P
S Z A U P U Z U Z M A D P Y B
I Y G H U B A R Y U P S G N X
J G X G P D A Y E L I Z T U O
U A I O G E K N B T Z Q I K Q
Y Y G R A K L Z U U T I X H A
R Q A I I R N E B A P X S A Y
N H P L J D B B H Q Q V K E M
I O L L A P H E M Z P Z S C K
I M I A G K B L Z D T M E F Y
V N H L S U A A B F P Y O F O
N X I T L X E Z N J B X I M C
S N N L M B N G I R A F F E V
Q A K F R J G B Z E C I E W L
```

elephant	oryx	lion
okapi	gorilla	zebra
giraffe	ostrich	

An **okapi** is a **large
mammal** from Africa
in the same family as
the giraffe. Look closely
at their diorama in the
**Akeley Hall of African
Mammals**. There's an
intruder! The artist added a
hidden chipmunk in the
background.

The **white rhinoceros** diorama has a **hidden animal** that most people miss. Can you guess what it is? Hint: It's pretty prickly. There's a **porcupine** bristling in fright at the charging rhinos! See if you can spot it.

When fossilized **animal bones** arrive at the Museum as part of a plaster matrix, they need to be cleaned for study and exhibit. People who prepare specimens are called **preparators**. There are different ways to prepare specimens. Many small fossil bones are **fragile**, so preparators have to be **very careful**. Sometimes they use a **small dentist's drill** to remove rock stuck to the fossils.

Some animal tissue
is **frozen** for future
research. The Museum
doesn't use freezers, though.
Small samples are placed
in **giant vats** filled with
a special chemical called
liquid nitrogen. This
chemical freezes everything
it touches and preserves
samples for a long time.

QUICK QUIZ

What do you call a scientist who studies mammals?

a) herpetologist

b) mammalogist

c) geologist

d) gemologist

Answer: b

Stop by the fourth floor to see the **first mastodon ever found** in the United States. It was discovered in the exact position it died in **11,000 years ago**. That's because it sank in a bog. The mud kept it well preserved.

The Museum gave the
**Bernard Family Hall of
North American Mammals**
a makeover. The project took over a
year and was completed in 2012.

They added
color to the faded animal fur,
dusted the leaves in the dioramas,
replaced cougar whiskers,
and updated the lighting to
make it look more natural.

★ ★ ★

This hall has 46 different types of mammals. One of them is the 9-banded armadillo. Many people consider the dioramas in this hall to be the best in the world!

Have you ever found a cool rock, seashell, or fossil that you think might be special? Each year, the Museum hosts an **Identification Day**. People are encouraged to bring in their treasures, and Museum scientists look over the specimens! The Museum's website **amnh.org** even has guidelines on what to bring.

In 2017, a woman brought a
bone to be identified. It turned out to
be a **Plesiosaur vertebra**!

★ ★ ★

These ancient reptiles lived
65 to 200 million years ago.
The woman and her husband were
shocked. They had been using
it as a paperweight!

A botanist who visited the Museum commented that the bison diorama had the wrong grass: it should have been buffalo grass. The Museum checked. Sure enough, the botanist was right! The Museum sent an expedition out West to find the right grass, and the diorama was fixed.

In the **Bernard Family Hall of North American Mammals**, there are **two** wolves on the hunt. They are chasing an **unseen deer** through the snow. But how do you make snow look real in a museum? It took a lot of creativity, but the artists came up with an idea. They used plastic foam for the snow and then added crushed marble dust to make it sparkle.

PEOPLE AND CULTURES

Anthropology is the study of **people and their cultures**. Scientists can find out more about people by studying the objects they used. These objects are called **artifacts**. The Museum has collected **hundreds of thousands** of them!

The Anthropology Division has almost **11 miles** of shelving to store **artifacts** in the Museum. It would be impossible to display them all! Luckily, you can still see them. When an object comes to the Museum, it is tagged with a number, and a photograph is put on the Museum's website.

Anthropology exhibitions are open to the public and appear in **six different buildings**! Storage takes up **four more buildings**.

The Museum is filled with **cool objects from around the world**, but anthropologists also collect other items that you might not think about. Anthropologists collect **photographs**, **letters**, **maps**, and **books**. They even make **sound and video recordings**.

The **Anthropology Division**
has built up a huge collection.

The division has
535,700 artifacts.

★ ★ ★

The Museum only displays
15,200 of them.

★ ★ ★

They add about **365**
artifacts every year.
(That's about **1 new
artifact each day**!)

Insects can invade wooden cases that hold items made with **fur**, **skins**, and **feathers**. To solve the problem, the Museum has switched from wooden cabinets to metal ones over the last 100 years.

QUICK
QUIZ

Large crates are stacked up against the walls in secret hallways throughout the Museum. Can you guess why?

a) Objects are always being shipped to other museums for study.

b) The Museum doesn't have enough space.

c) The crates look nice.

d) The crates are used as desks.

Answer: a

In Vietnam, family members burn paper objects at funerals. The objects are made to look like items the dead person cared about during life. When a scientist from the Museum saw a **life-size paper bicycle** being made for a funeral, she asked if one could be made especially for the Museum. It was brought back to the museum in 2003 and is kept with the collections on the fifth floor.

Most gold and silver artifacts from ancient times were melted down long ago. Luckily, a **500-year-old royal llama** made of silver from the Inca civilization was found in great condition. It is from the Island of the Sun in Bolivia. Scientists think a wealthy ruler once owned it. Look for it in the **Hall of South American Peoples**.

The **3,000-year-old** Kunz Axe is made of a stone called **jade**. This **10-inch-long** carving portrays a figure that is half human and half jaguar! The axe is one of the Museum's most important objects from ancient Mexico. That's partly because it's the **biggest piece of jade ever found** in Mexico!

QUICK QUIZ

One of the best ways to find out about people is to study their trash. That's exactly what scientists do to find clues about past civilizations. Can you guess what ancient trash heaps are called?

a) garbage heaps

b) middens

c) dump grounds

d) stinky hills

Answer: b

Look for **pigeon whistles** in the **Stout Hall of Asian Peoples**. These are paper-thin whistles made of bamboo. People used to strap them to the backs of pigeons. As the birds flew, the wind caused the whistles to **blow a tune**. Behind the scenes at the Museum, broken pigeon whistles are important, too. They help scientists see how the whistles were made.

Need a place to keep your pet cricket? How about a cricket gourd? These are **ancient Chinese containers made from a large piece of fruit**. Once the fruit is dried out, it is decorated with materials such as **wood** and **ivory**. Then it's ready for a cricket. The insects were sometimes used in cricket fights. You can find the gourds in the **Stout Hall of Asian Peoples**.

One of the Museum's most popular artifacts is the **Great Canoe**. At **63 feet long**, this Native American canoe is the **largest of its kind**. It was made from a single cedar tree in **1878**. Find the killer whale painted on the prow. It's hiding in plain sight!

Some ancient robes and blankets are so delicate, they can never be folded. Instead, they are stored flat in **giant drawers** in cabinets. The largest cabinet is **6 feet wide** and **10 feet deep**! Once these fabrics are on the trays, they are rarely moved. That's because the elevators aren't big enough for the trays!

The specimens in the **Spitzer Hall of Human Origins** aren't real. They are casts! Casts are models based on **the real thing**.

Ever wonder how scientists came up with a name for a 3.18-million-year-old skeleton? When scientists were celebrating their discovery of "Lucy," a cast of which is on display in the Spitzer Hall of Human Origins, they were listening to a **famous Beatles song**. Can you guess which one? It was *Lucy in the Sky with Diamonds*! That's how Lucy got her name.

MYSTERIES
OF THE
NIGHT
SKY

When the **Hayden Planetarium** first opened, it didn't have fancy equipment. To create the night sky, scientists took a **copper plate** and hand-punched **tiny holes** in it. They projected light through the holes, and when the light reflected on the ceiling, it looked like stars!

QUICK
QUIZ

The Hayden Planetarium opened in 1935. Can you guess how many planetariums there were in the whole country at that time?

a) three

b) four

c) eight

d) ten

Answer: b

Today the **Zeiss Mark IX projector** does the job of the old copper plates with the hand-punched holes. This machine took **two** years to design. The projector weighs **8,000 pounds** and has **30** different motors. It's so big, it had to be shipped to the Museum in **14** separate crates.

In **2000**, the Museum opened the rebuilt Hayden Planetarium in a new building called the **Rose Center for Earth and Space**.

Years it took to rebuild the planetarium: **6**

★ ★ ★

Number of visitors **per hour** for the first 18 months: **1,000**. That's over 2.5 million visitors!

The plans for the Rose Center started out as a **drawing on a napkin**! The Museum was getting ready to fix up the old planetarium. During a meeting, the architect grabbed a napkin and began to sketch. The Museum officials liked it so much, they used that drawing as a plan for a **whole new building**. When it was finally complete, it looked just like the drawing.

The planetarium has a one-of-a-kind design: it's a giant sphere suspended inside a glass cube!

Height of the sphere: **87** feet

★ ★ ★

Height of the cube:
95 feet

★ ★ ★

Number of seats in the planetarium dome: **432**

The Rose Center's cube is made of a **special type of glass**. It's called **water white.** The glass is almost completely clear. When the light hits it a certain way, it doesn't look like the glass is there at all!

The Hayden Planetarium has had a lot of visitors over the years, but none are as famous as **Superman**. DC Comics teamed up with the Museum to make a comic strip about the superhero's home planet, Krypton. Neil deGrasse Tyson, Frederick P. Rose Director of the Hayden Planetarium, wanted to find a real star in the night sky that Krypton might orbit around. He used facts about Superman's life to help him. Together with DC Comics, they picked a star called **Corvus**.

Wonder what it's like to **travel through time and space**? After visiting the **Big Bang Theater**, follow the **360-foot Heilbrunn Cosmic Pathway** around the sphere at the Rose Center for Earth and Space. Each step will take you back **75 million** years. Once you've walked around it, you'll have traveled through **13 billion** years of space history!

On the Heilbrunn Cosmic Pathway, different objects represent different amounts of time. At the end is a tiny surprise—a single human hair! It represents the past 30,000 years of human history, which isn't that long since the universe has been around for 13 billion years.

Word Search

Ready, Set, Blast Off! Search for the words below to see what you'll find at the Rose Center. (Words can be found vertically, horizontally, and diagonally.)

```
O D E M S B G S P H E R E Z C
B I R U T B A P A T H W A Y N
K J I I A U L R N G Y C S E T
K L L R R J A L E T O T G K H
U K S A S H X L S T E I W F U
C E X T F O I S Z N A H Q Z Y
M G M E I S E K A T R E M D U
Z J F N C V S L S P T J H Z Q
J L W A O W P K F H N W M T U
Q B R L E U W R E X K Q T E W
P Y M P L P C Y P S E T Q R J
E L K E A K C H D X K F E Q V
R X J K U E S P A C E L H I B
V M Z S M K D V Q J U E T Q M
M E T E O R I T E A S T J U P
```

galaxies   space   planetarium

planets   pathway   stars

meteorite   sphere
```
```

The Willamette Meteorite is the largest meteorite ever found in the United States and the sixth largest meteorite in the world! Scientists think it was part of a planet that exploded billions of years ago. When it blasted through Earth's atmosphere, it began to melt. As it cooled, large cavities from bubbles formed. Check it out in the Cullman Hall of the Universe.

The Willamette Meteorite was found within the Upper Willamette Valley of Oregon near the present-day city of Portland. This is where many **Clackamas Indians** live. The Clackamas named the meteorite "Tomanowos." In Clackama tradition, Tomanowos is a spiritual being that has healed and empowered the people of the valley since the beginning of time.

QUICK
QUIZ

Guess how much the Willamette Meteorite weighs.

a) 200 pounds

b) 12,500 pounds

c) 31,000 pounds

d) 1 million pounds

Answer: c

The Ahnighito Meteorite (pronounced ah-ni-GEE-to) weighs almost 70,000 pounds. The meteorite is so heavy that it has supports called "footings" that go right through the first floor of the Museum and continue all the way down into the basement. They keep on going right into the bedrock of the Earth! Check it out in the Ross Hall of Meteorites. The 4.5-billion-year old Ahnighito is almost as old as the sun!

Want to explore the universe? The American Museum of Natural History's **Digital Universe** is an amazing 3-D atlas that can take you from our solar system to the edge of the observable universe. You can download a free version of this amazing 3-D atlas from the Museum's website to use at home!

In **1992**, thousands of people were watching a high school football game when a **giant rock fell from the sky**! It crashed into a parked car in the town of Peekskill, New York. Many people caught the meteorite's fall on video. Today, the rock is in the **Ross Hall of Meteorites**. Look closely and you'll see that the red paint from the car is still there!

The **smallest** item on exhibit in the Museum is a vial of dust. **Stardust,** that is! As stars age, they release grains of dust into space. This vial of stardust is made up of **60 quintillion** tiny diamonds! That's a **6** with **16** zeros after it! This stardust hitched a ride to Earth on the Allende Meteorite. Find the vial in the **Ross Hall of Meteorites**.

ANSWER KEY

PAGE 61

```
R C B A C W C V J F H H A I Z
P I Z J N V A G U Y M O A V J
O S F E L J M N M C Q D X K J
S H O V E L E G O B K P B Z G
G V X B H R R U O D C K F J H
M O B I I E A O S X D P L E J
Z E L D Z P E H S X X F Z F
U L X U F S H X O E Y R D C U
G L O V E S A Y Z F X S L G G
N U M Z X K M X P I H S Y R P
H A V P N U M T S N U N F P I
P H C A V L E Q Z K O L J X T
H C F K X Q R J T G E S B N B
F G Z S E E T K W B P X S Q O
Z P P B P R C X I W T N B G X
```

PAGE 100

```
T V J G Y Z Z D F Q W M K K S
R Y D W I X L J V R L P P I Z
A U Q S N E I D Z Z B S N G E
X J J A E R M K X U E I L F S
V Q Y Y G W O B F S X W F R T
S C Y L U S X H L M V B E B C
P R E H I S T O R I C H O O N
U J K C S E T X Y T T I S N I
G V G H M P U K I A U G H E T
D G Q H X R G U E S U S F E X
K D O E C L Y F P Z C L U Q E
R H G L A Z Z Q J Y E X G T T
L G A R S V T U J T N B Y Y I
S W T Y A G B L I S S O F B Y
S A K G U E R G P S Q C F Q P
```

PAGE 107

```
T O S T R I C H A F O O D S Y
J N Z K J U P N H K K G R W P
S Z A U P U Z U Z M A D P Y B
I Y G H U B A R Y U P S G N X
J G X G P D A Y E L I Z T U O
U A I O G E K N B T Z Q I K Q
Y Y G R A K L Z U U T I X H A
R Q A I I R N E B A P X S A Y
N H P L J D B B H Q Q V K E M
I O L L A P H E M Z P Z S C K
I M I A G K B L Z D T M E F Y
V N H L S U A A B F P Y O F O
N X I T L X E Z N J B X I M C
S N N L M B N G I R A F F E V
Q A K F R J G B Z E C I E W L
```

PAGE 149

```
O D E M S B G S P H E R E Z C
B I R U T B A P A T H W A Y E
K J I I A U L R N G Y C S E T
K L L R R J A L E T O T G K H
U K S A S H X L S T E I W F U
C E X T F O I S Z N A H Q Z Y
M G M E I S E K A T R E M D U
Z J F N C V S L S P T J H Z Q
J L W A O W P K F H N W M T U
Q B R L E U W R E X K Q T E W
P Y M P L P C Y P S E T Q R J
E L K E A K C H D X K F E Q V
R X J K U E S P A C E L H I B
V M Z S M K D V Q J U E T Q M
M E T E O R I T E A S T J U P
```

LIST OF HALL NAMES

Akeley Hall of African Mammals

Anne and Bernard Spitzer Hall of Human Origins

Arthur Ross Hall of Meteorites

Black Hole Theater

David H. Koch Dinosaur Wing; consists of the Hall of Ornithischian Dinosaurs and the Hall of Saurischian Dinosaurs.

David S. and Ruth L. Gottesman Hall of Planet Earth

Discovery Room

Dorothy and Lewis B. Cullman Hall of the Universe

Felix M. Warburg Hall of New York State Environment

Grand Gallery

Gardner D. Stout Hall of Asian Peoples

Hall of African Peoples

Hall of Asian Mammals

Hall of Biodiversity

Hall of Birds of the World

Hall of Eastern Woodlands Indians

Hall of Mexico and Central America

Hall of North American Forests

Northwest Coast Hall

Hall of New York City Birds

Hall of New York State Mammals

Hall of Plains Indians

Hall of Primates

Hall of Primitive Mammals; part of the Lila Acheson Wallace Wing of Mammals and Their Extinct Relatives

Hall of Reptiles and Amphibians

Hall of Small Mammals

Hall of South American Peoples

Hall of Vertebrate Origins

Harriet and Robert Heilbrunn Cosmic Pathway

Hayden Planetarium

Hayden Big Bang Theater

Jill and Lewis Bernard Family Hall of North American Mammals.

LeFrak Family Gallery

Leonard C. Sanford Hall of North American Birds

Margaret Mead Hall of Pacific Peoples

Irma and Paul Milstein Family Hall of Ocean Life

Paul and Irma Milstein Hall of Advanced Mammals; part of the Lila Acheson Wallace Wing of Mammals and Their Extinct Relatives

Scales of the Universe

Theodore Roosevelt Memorial Hall

Theodore Roosevelt Rotunda

Wallach Orientation Center

Whitney Memorial Hall of Oceanic Birds

We hope you enjoyed
this brief tour through the
secrets and highlights of
the American Museum
of Natural History. While
we've packed the book with
information, there is so much
more for you to discover.
So start exploring and find
secrets of your own!

If you enjoyed *Secrets of the American Museum of Natural History*, check out *Secrets of Our Nation's Capital, Secrets of the National Parks, Secrets of Disneyland,* and *Secrets of Walt Disney World*!